COUNTDOWN TO SPACE

SPACE LAUNCH DISASTER
When Liftoff Goes Wrong

Michael D. Cole

Series Advisor:
John E. McLeaish
Chief, Public Information Office, retired,
NASA Johnson Space Center

Enslow Publishers, Inc.

40 Industrial Road   PO Box 38
Box 398   Aldershot
Berkeley Heights, NJ 07922 Hants GU12 6BP
USA   UK

http://www.enslow.com

Library of Congress Cataloging-in-Publication Data

Cole, Michael D.
    Space launch disaster : when liftoff goes wrong / Michael D. Cole.
        p. cm. — (Countdown to space)
    Includes bibliographical references and index.
    Summary: Describes the dangers of launching spacecraft, covering the Russian
Mars space probe explosion, the Apollo 1 fire, the Apollo 12 lightning strike, and
the Challenger explosion.
    ISBN 0-7660-1309-X
    1. Space vehicle accidents Juvenile literature. [1. Space vehicle accidents.]
I. Title. II. Series.
TL867.C6522   2000
363.12'4—dc21   99-32179
                                                                CIP

Printed in the United States of America

10 9 8 7 6 5 4 3 2 1

To Our Readers: All Internet addresses in this book were active and appropriate
when we went to press. Any comments or suggestions can be sent by e-mail to
Comments@enslow.com or to the address on the back cover.

Photo Credits: National Aeronautics and Space Administration, pp. 4, 9,
11, 12, 14, 16, 18, 20, 23, 25, 28, 29, 30, 32, 34, 36, 38, 39, 40;
© VideoCosmos, p. 7.

Cover Illustration: NASA (foreground); Raghvendra Sahai and John
Trauger (JPL), the WFPC2 science team, NASA, and AURA/STScI
(background).

Cover photo: The burned remains of the Apollo 1 *command module.*

CONTENTS

1 Dangerous Steps into Space 5

2 The Launch 8

3 The Apollo 1 Fire 17

4 Apollo 12 27

5 The Challenger Explosion 33

Chapter Notes 42

Glossary 44

Further Reading 46

Index 47

The Apollo 1 *capsule is lifted to the top of the gantry for its mission in 1967. It was the site of America's first space tragedy. But years prior to the Apollo 1 disaster, the Russians had a disaster of their own. . . .*

1

Dangerous Steps into Space

A mighty R-16 rocket sat on the launchpad at the Soviet space program's launch facility in Baykonur. Atop the rocket on this day, in October 1960, was a space probe designed to travel to the planet Mars. The many engineers and technicians at the launch facility watched as the seconds ticked through the final moments of the countdown.

Five . . . four . . . three . . . two . . . ONE.

Nothing happened.

The rocket's engines had failed to ignite. Instead of lifting off, the rocket sat there, motionless on the pad.

The situation was extremely dangerous. The rocket was fully loaded with hundreds of tons of highly explosive fuel. Until that fuel was pumped out of the rocket's tanks, there was great danger that the rocket could explode on the pad. Normal safety procedure stated that the people at the launchpad were supposed to pump away the fuel.[1]

But Soviet leader Nikita Khrushchev wanted the rocket launched as soon as possible. Pumping the fuel away and restarting the countdown could result in delaying the launch by days.

The Soviet rocket engineers were ordered to ignore standard procedure. Soon a number of engineers and technicians were approaching the fully fueled rocket to see why it had not fired.

Suddenly there was an explosion at the base of the rocket. The explosion blew apart the fuel tanks, throwing tons of fuel into the air. The fuel caught fire and the flames spread rapidly. It was too late for anyone to escape. The entire pad burst into a tremendous explosion that destroyed everything and everyone for hundreds of feet in every direction around the launchpad.

Nearly one hundred people were killed, including a number of leading engineers in the Soviet space program. The accident was a tragic setback.[2] Soviet leaders would not be sending a probe to Mars. They would also not be sending a person into space as soon as they thought.

No cosmonauts were involved in the mission that day. But because so many people were killed at the launchpad, the accident at Baykonur in October 1960

The Soviet launchpad explodes after a failed launch of a space probe. Firefighters struggle to put the fire out (inset).

ranks as one of the worst tragedies in the history of spaceflight.

Spaceflight is a difficult and dangerous operation. As humans began their first journeys into space in the 1960s, the scientists and engineers for the United States and Soviet space programs had much to learn. Those lessons were sometimes learned the hard way, at the cost of human lives.

Since 1957, the world's space programs have been sending vehicles into space. Yet after forty years of spaceflight experience, the hardest and most dangerous task is still the launch.

2

The Launch

A tremendous amount of energy is required to get astronauts and their spacecraft off the ground and into space. The average jet airliner travels at speeds of about 450 miles per hour. However, a speed of about 25,000 miles per hour must be achieved in order for a spacecraft to break away from Earth's gravity.

A number of different rockets have carried American astronauts into space. Redstone rockets carried Alan Shepard and Gus Grissom into space in 1961. John Glenn became the first American to orbit Earth when an Atlas rocket boosted his capsule, *Friendship 7*, into orbit in 1962. Atlas rockets carried three more Mercury astronauts into orbit in 1962 and 1963. Titan rockets were used to launch the two-person Gemini capsules into orbit. The Saturn V, the largest and most powerful rocket ever built, launched the Apollo spacecraft on their missions to the Moon. Today, a combined system of fueled rockets and solid rocket boosters are used to launch the space shuttle into space.

All these rockets were designed to do the same thing. They must lift the spacecraft off the ground and get it to a speed that will carry it into space. The phase of the flight that gets the spacecraft into space is called the launch.

Launch is the most dangerous phase of a space-flight because the rocket is filled with many thousands of gallons of explosive fuel. If

In the early 1960s, the Atlas rocket blasted John Glenn and other Mercury astronauts into space.

anything goes wrong on the launchpad or during the launch, the entire vehicle and its astronauts could be destroyed in a gigantic explosion.

An emergency such as this happened during the attempted launch of *Gemini 6* on December 12, 1965. Atop the rocket in their Gemini spacecraft, astronauts Wally Schirra and Tom Stafford listened on their headsets as the countdown neared its end.

"You are cleared for takeoff," said the capsule communicator, or capcom.

A voice from the blockhouse near the launchpad counted, "Ten . . . nine . . . eight . . . seven . . . six . . . five . . . four . . . three . . . two . . . one . . . ignition."

The rocket engines began to roar, but only 1.2 seconds later, they stopped.

"We have a shutdown, *Gemini 6*!" the astronauts heard in their helmets. Schirra glanced at the mission clock in the spacecraft. It had started. But the rocket was going nowhere. Schirra and Stafford were sitting atop what at any moment could become an exploding bomb.[1]

Everything Schirra had learned in the mission rule book told him he should reach down and pull the ring that would shoot Stafford and himself out and away from the capsule in their ejection seats. But Schirra waited, not pulling the ring.

Another Gemini spacecraft, *Gemini 7*, was already in space waiting to rendezvous in orbit with *Gemini 6*. Schirra knew that firing the ejection seats away from the

One second after the rockets roared, the Gemini 6 *launch shut down. Astronauts Wally Schirra and Tom Stafford sat on top of the rocket that could have exploded at any time.*

capsule would make it impossible to get the spacecraft ready again in time for its rendezvous with *Gemini 7*. *Gemini 7* had already been in space for days. It would have to come down before new ejection seats could be put back in *Gemini 6* to make it ready for launch again. This was true, assuming the entire rocket did not explode underneath them.

If it *did* explode, and Schirra and Stafford did not eject in time, they would be killed. Hoping to save the mission for another day, Schirra gambled that the rocket was still safe.

"Okay," he said, "we're just sitting here breathing."[2]

Schirra had gambled correctly. A stuck fuel valve had

Astronaut Wally Schirra sat waiting in the Gemini 6 capsule. After the rocket's fuel was pumped away, the astronauts left the spacecraft safely.

caused the engines to automatically shut down. The one hundred thirty-two tons of fuel were pumped away, and the astronauts were removed from the spacecraft. This time the disaster was avoided, allowing Schirra and Stafford to make their rendezvous with *Gemini 7* three days later.

The danger of explosion during a launch is so great because the launch itself is actually a type of controlled explosion. Tons of rocket fuel are carefully pumped, mixed, and ignited. The process creates a continuous explosion, or combustion, of the rocket fuel. The combustion from the rocket engines pushes the rocket higher and faster toward space.

Today's space shuttle is made up of four parts called the space shuttle launch assembly. These parts are the space shuttle orbiter, the external fuel tank, and two solid rocket boosters, or SRBs.

The space shuttle orbiter is the winged spacecraft that carries the astronauts into space and returns them to Earth. It has a total of five engines. Three are used to lift the shuttle off the pad and shoot it rapidly into space. The remaining two, Orbital Maneuvering System engines, or OMS engines, are used near the end of the launch to raise the shuttle's orbit.

The orbiter is attached to a large external fuel tank. This tank, shaped like a giant thermos bottle, contains 1.5 million pounds of liquid hydrogen and oxygen. These chemicals are mixed together and burned in the shuttle

The winged shuttle orbiter is attached to one large external fuel tank. The two smaller solid rocket boosters are located on each side of the external fuel tank.

orbiter's engines during the launch.[3]

The two solid rocket boosters are arranged on either side of the external fuel tank. These boosters burn solid propellants during the first two and a half minutes of the launch. Each solid rocket booster is as tall as a fifteen-story building and produces 2.6 million pounds of thrust, equal to the power of more than thirty thousand locomotives.[4]

Launch of the space shuttle begins as the shuttle's main engines are started near the end of the countdown. The fuel in the external tank fuels these main engines. Next, the solid rocket boosters are ignited. While the shuttle is still clamped to the launchpad, these engines build up to the proper thrust for liftoff.

Shortly after the countdown reaches zero, both the main engines and solid rocket boosters reach the proper thrust, and the hold-down clamps release. The shuttle then lifts up and away from the pad. Bright flames gush from the rockets, and a column of thick white smoke is left behind as the shuttle thunders into the sky.

The shuttle's velocity increases rapidly. Two minutes into the launch, the shuttle is already traveling faster than the speed of sound, which is 740 miles per hour. The shuttle's pilot throttles up the main engine, increasing the shuttle's velocity even more. Seconds after throttle-up, the two solid rocket boosters are dropped away from the sides of the external tank. The boosters are now empty. They drift by parachute down to the Atlantic Ocean, where they are recovered and reused in later missions. The main engines burn for another six minutes, carrying the shuttle higher and faster toward space.

Eight and a half minutes after leaving the pad, the shuttle experiences MECO—main engine cutoff. It is now traveling at more than seventeen thousand miles per hour, at an altitude of more than one hundred miles above Earth.

The external fuel tank then separates from the shuttle and burns up in Earth's atmosphere. Final thrusts from the shuttle's OMS engines push the orbiter to a higher and more circular orbit, about one hundred fifty miles above Earth. The launch is now completed, and the shuttle is successfully in orbit.[5]

Somewhat like the shuttle launch system, the mighty Saturn V rocket carried the Apollo spacecraft into space in stages. Each of these stages used liquid fueled rockets. The towering 320-foot-tall Saturn V was the largest and most powerful rocket ever built. It had to be to shoot three astronauts and their spacecraft toward the Moon.

Launching a spacecraft is an amazingly complex operation. Yet in only twelve years, the world's space programs had gone from launching the first satellite in 1957 to launching three astronauts to the Moon in 1969.

Launching so many missions into space was an impressive achievement. Scientists and engineers had overcome many technical problems in the designing, building, and launching of these spacecraft.

Despite the many successes, and the skillful design of the various spacecraft, these launches did not always go perfectly. The Apollo program, in fact, began with a terrible tragedy.

The mighty Saturn V rocket that lifted the Apollo 11 capsule into space was the largest and most powerful rocket ever built.

3

The Apollo I Fire

NASA had experienced much success with the Mercury and Gemini manned space vehicles. By 1967, NASA was ready to test the Apollo spacecraft that would carry astronauts to the Moon.

Astronauts Gus Grissom, Ed White, and Roger Chaffee were chosen to make the first Apollo flight. The three astronauts were to test the new spacecraft by making a number of orbits around Earth.

The Apollo spacecraft had experienced a number of problems during its development. Even as it was placed atop the Saturn 1B rocket and taken to the launchpad, some questions and concerns about its design remained. Despite these concerns, the program and launch schedule went forward for a number of reasons.

President John F. Kennedy had made a speech in 1961, in which he said the United States should land an astronaut on the Moon before the end of the decade. The leaders at NASA took President Kennedy's words very seriously and worked hard to achieve the goal the president had set. NASA also wanted to reach the Moon first, before the Soviet space program had a chance to get there. Getting to the Moon first would be an impressive accomplishment for the country that did it.

Those pressures pushed the Apollo program ahead, perhaps too quickly. NASA was working very rapidly to develop and launch the extremely complex Apollo

Astronauts Gus Grissom, Ed White, and Roger Chaffee were chosen to make the first Apollo flight.

spacecraft. At such a pace, it was difficult to maintain the safety levels that a manned spacecraft required.[1]

The scheduled launch of *Apollo 1* was three weeks away when the astronauts suited up for a full rehearsal of the launch. It was January 27, 1967. Grissom, White, and Chaffee wore their full space suits and helmets. A van took them to the launchpad just as it would on the day of the launch. Then they rode the elevator to the Apollo spacecraft, where it awaited them atop the Saturn 1B rocket. By 3:00 P.M., the astronauts were inside the spacecraft and the practice countdown began.

This launch rehearsal was not considered to be a dangerous test. Fuel had not yet been pumped into the rocket, so there seemed to be little danger of fire or explosion. But everything else about the rehearsal was just as it would be on the day of launch. Once the three astronauts had strapped themselves to their couches, the spacecraft cabin was filled with pure oxygen. The launch rehearsal was done, using the spacecraft's internal electrical power.

To keep the pure oxygen atmosphere in the cabin, the spacecraft's hatch was securely locked and sealed. With the hatch sealed, any attempt by the astronauts to escape from the cabin in case of emergency would take at least ninety seconds.[2]

For the next three hours, a series of problems interrupted the practice countdown and troubled the astronauts. It bothered Grissom that at this late stage in

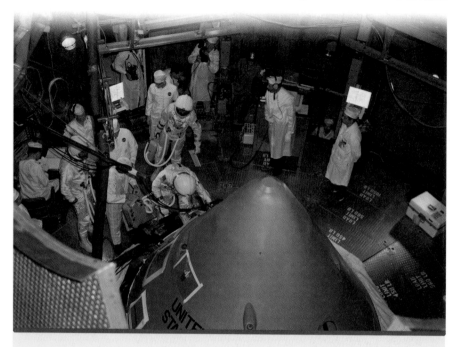

Grissom, White, and Chaffee enter the Apollo 1 *capsule for a launch rehearsal. They were preparing for the first launch of an Apollo spacecraft.*

their preparations to launch Apollo, there were still problems with the communications system.[3] As the series of malfunctions continued, Grissom grew impatient.

"If I can't talk with you only five miles away, how can we talk to you from the Moon?" he snapped.[4]

The practice countdown was stopped at T minus ten minutes. It was already 6:20 P.M. and had grown darker around the launch area. The people in the blockhouse near the launchpad, as well as the technicians in the white room area attached to the spacecraft, were getting weary of the long and problem-filled launch rehearsal.[5]

Instead of getting better, the rehearsal seemed to be getting worse.

"I can't even shut off my microphone now!" Grissom complained.

"Let's cancel out today," said an engineer to one of the launch controllers. "This could go on forever. We're better off if we shut down and do the full test again."[6]

Despite the difficulties, the launch controllers decided to continue the countdown minutes later. At 6:31 P.M. there was a momentary electrical spark somewhere in the twelve miles of electrical wiring running throughout the spacecraft. Fire needs oxygen in order to burn. The spacecraft cabin was full of pure oxygen. On contact with the rich oxygen inside the spacecraft, the spark exploded into flame.

"Fire, I smell fire," said Roger Chaffee almost calmly.[7]

The shock wave of flames then engulfed the cockpit in an instant. The astronauts immediately squirmed out of their couches to try making an emergency escape. But it was already too late. The inside of the spacecraft was an inferno of death in a matter of seconds.

"Fire! We've got a fire in the cockpit!" Grissom yelled urgently.[8]

"We've got a bad fire . . . we're burning up here!" Chaffee cried. They were the last understandable words from the spacecraft.[9]

Within twenty seconds, the heat and pressure inside the cabin became so great that the outside wall of the spacecraft split. Fire, smoke, and fumes filled the white room surrounding the spacecraft. The technicians inside the white room barely escaped being killed from the explosion and toxic fumes. They grabbed fire extinguishers and did everything they could to put out the fire until the firefighters arrived with more equipment.

When the fire was out and the smoke cleared, there was no hope that anyone inside the spacecraft could have survived the fire and explosion. Deke Slayton, Grissom's friend and a fellow astronaut during the Mercury program, was in the blockhouse when the fire occurred. He and others raced out of the blockhouse toward the pad to see what had happened. When he reached the white room, his worst fears were realized.

"It was devastating," Slayton said. "Everything inside was burned, black with ash. It was a death chamber."

The three men had not been burned to death. They had suffocated in the fatal mix of toxic fumes released when the interior hoses, switches, wires, and all other instruments inside the cabin began to burn. Within moments, there had been nothing to breathe in the flame-swept cabin but deadly fumes.

"The suits had protected them from the flames," Slayton added. "None of them had any physical burns. . . . It was the . . . environmental control systems that got them. . . . I had to turn my head."[10]

Even as the deadly inferno swept the spacecraft, the three astronauts performed as they had been trained.

The bodies of Gus Grissom and Roger Chaffee were found crumpled on top of Ed White's. Grissom had been trying to push against the lever that would instantly depressurize the cabin. This was the first step in getting the hatch open for an escape. Grissom tried, but within seconds the lever itself would have been in flames.

Ed White had reached up with his left hand to help Grissom try to release the inner hatch. Chaffee had remained in his couch for a few moments, trying to maintain contact with the controllers. As the flames rolled up and around him, he made a final plea for the crew to get them out. He then reached over across Ed

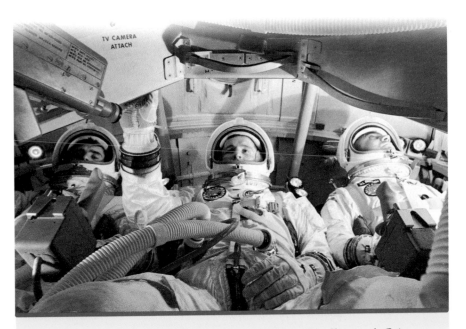

During the practice session, astronauts White, Chaffee, and Grissom were seated in the Apollo 1 *capsule, where the deadly fire occurred.*

White in a desperate effort to help Grissom with the hatch release mechanism.

Within seconds the flames roaring in front of their helmet visors would have made it impossible for Grissom, White, and Chaffee to see what they were doing. Just a few moments later, there was no more breathable air in the cabin.[11]

There had been no attempt to launch. The rocket was not even fueled. The crew of *Apollo 1* had never gotten off the ground. Yet on Launchpad 34 at Cape Kennedy, the grim fact was that astronauts Gus Grissom, Ed White, and Roger Chaffee were dead.

The *Apollo 1* fire shocked NASA and the nation. In the months following the tragedy, a full investigation was made of the accident. Investigators found that the spark had come from a set of wires that had lost their insulation and become exposed. The spark most likely occurred right below Gus Grissom's couch. The resulting flames then spread throughout the spacecraft.

The investigation's report criticized other safety flaws in the Apollo spacecraft's design. NASA never tried to cover up these flaws. Instead, the space program got to the bottom of these problems and solved them. One of the first changes the report recommended was that the pure oxygen atmosphere never be used again inside the spacecraft during ground simulations.

The tragedy had caused an understandable change in the people at NASA. The engineers and technicians

The burned remains of the Apollo 1 command module and the fire-damaged interior (inset) show the effects from the intense heat of the fire.

involved with the Apollo program were now very committed to seeing that the three astronauts did not die for nothing. These efforts resulted in a safer redesign of the Apollo spacecraft.[12]

Nearly two years passed during the *Apollo 1* investigation and the redesigning effort. NASA had lost twenty months in its race to reach the Moon before the end of the 1960s. But the accident had taught the people at NASA not to push recklessly beyond the safety requirements of their technology.[13] *Apollo 2* through *Apollo 6* were a series of unmanned test flights that showed both the rocket and the redesigned Apollo spacecraft were ready to go into orbit.

On October 11, 1968, *Apollo 7* lifted off from Cape Canaveral with the crew of Wally Schirra, Donn Eisele, and Walter Cunningham. The flight went almost perfectly. No difficulties or system failures in the new Apollo spacecraft occurred during the entire mission.

The success of *Apollo 7* and the redesigned Apollo spacecraft put NASA back on schedule for landing an astronaut crew on the Moon in 1969. America's space program had learned the lessons needed to get Apollo spacecraft safely and successfully into space. But those lessons had been learned at a high price.

It had cost three astronauts their lives.

4

Apollo 12

Apollo 7, Apollo 8, Apollo 9, and *Apollo 10* prepared NASA for its first mission to land on the Moon.

Apollo 11 was a triumph! On July 20, 1969, Neil Armstrong and Edwin "Buzz" Aldrin became the first human beings to walk on the Moon. Their historic mission to the Moon had gone almost perfectly. Things did not go as smoothly for the next mission, *Apollo 12*.

The crew of *Apollo 12* was commander Charles "Pete" Conrad, lunar module pilot Alan Bean, and command module pilot Richard Gordon. Conrad had flown in space already on two Gemini missions. Gordon had been Conrad's partner on the *Gemini 11* mission. *Apollo 12* would be Bean's first spaceflight.

The second manned mission to the Moon was

Charles Conrad, Richard Gordon, and Alan Bean were chosen as the astronauts to fly the second manned mission to the Moon—Apollo 12.

scheduled for liftoff on the morning of November 14, 1969. As the time for liftoff drew near, the weather around the launch area grew poor. Dark clouds gathered, and it began to rain. Conrad waited, strapped in the commander's couch inside the spacecraft. He looked out his window and saw rain droplets seeping inside the protective covering of the nose of the spacecraft. It had to be raining hard for the water to seep inside the covering. He knew a launch in such weather would be rough.[1] But there were pressures not to postpone the mission.

Instead of getting better, the weather around southern Florida for the next few days was expected to get worse. If the launch was canceled that morning, the expected weather system might stretch the launch delay to several days. NASA did not want to disappoint important supporters of their program, including President Richard Nixon. The launch was given a "Go" despite the weather.

At 11:22 A.M. *Apollo 12* lifted off the launchpad. Within twenty seconds, the giant Saturn V rocket disappeared into the low-hanging clouds. The rocket and its long trail of flame ripped through the gray sky of clouds and rain. As it vanished into the clouds, a bolt of lightning struck the launch complex area.

"What's happening to the Saturn V?" someone in the spectator area cried out.[2]

Lightning was not only flashing near the launch area. Streaking through the thick rain clouds, *Apollo 12*'s metal surfaces were building up static electricity that could cause problems for the electrical systems aboard the spacecraft. The crew saw a bright flash and heard a loud boom outside the spacecraft. An instant later, warning lights lit up their instrument panels. There were so many lights the three astronauts could not read them all. Sixteen seconds later, there was another flash and boom.

Suddenly the spacecraft went dark. For a few seconds the astronauts could see and do nothing. Then the backup batteries kicked on. The cabin lights flashed back on, and power came back to

A bolt of lightning hits the launch area during the liftoff of Apollo 12.

Apollo 12 *lifted off during a thunderstorm. The astronauts on board were worried they might have been hit by lightning.*

the spacecraft. Now the instrument panel was blinking with red warning lights everywhere! With surprising calm, Conrad reported to Mission Control.

"We just lost the guidance platform, gang," he said, referring to the computer system that controlled the spacecraft.[3] The astronauts worked quickly to restore order to the systems in their spacecraft.

"We are weeding out our problems here," Conrad said. "I don't know what happened. I'm not sure we didn't get hit by lightning."[4]

Although some sort of electrical disturbance seemed to have occurred in their spacecraft, the rocket continued to function properly. They continued to rocket toward orbit.

"Got a little vibration of some kind," Gordon said of the rocket, "but she's chugging along here minding her own business."

Gordon punched the switches on some circuit

breakers to put the spacecraft back on its main electrical system. Everything seemed to be working.[5]

Minutes later *Apollo 12* achieved orbit around Earth. The astronauts checked the command module and the lunar module, getting them ready for the trip to the Moon. They also made sure the two spacecraft had not been damaged during launch. As they worked, Conrad discussed the bumpy launch with Mission Control.

"Your theory that it was probably lightning that did it, that looks about the best idea," said Mission Control.

"We were a pretty big piece of electricity builder going through there," Conrad replied, "so we might just have discharged ourselves."[6]

Apollo 12 may have been struck by lightning. It is also possible that the spacecraft discharged its own static electricity, built up by traveling so rapidly through the storm's turbulent air. What is certain is that the three astronauts were lucky. If the lightning or static discharge had affected the Saturn V rocket's guidance systems instead of the Apollo spacecraft's electrical systems, the astronauts would have been sent rocketing out of control over the Atlantic Ocean.[7] They would have been forced to trigger the escape mechanism, firing the Apollo capsule away from the Saturn V. If the rocket had been too badly affected by the electric charge, the astronauts might even have been unable to abort. They could easily have been killed.

The astronauts aboard *Apollo 12* were able to

After a rocky start, the three Apollo 12 *astronauts completed their Moon mission and returned safely to Earth. Their capsule splashed down in the ocean, and they were recovered by the United States Navy.*

continue their mission and make a successful landing on the Moon. All three astronauts returned safely to Earth.

The next mission, *Apollo 13*, experienced an explosion aboard the spacecraft on the way to the Moon. The explosion caused a loss of power that caused the Moon landing to be canceled. The crew barely made it home alive.

The launch was about the only thing that went right with *Apollo 13*. With the successful missions of the rest of the Apollo program in the 1970s, and the beginning of space shuttle flights in the 1980s, launching astronauts into space began to seem routine. It was not. America's greatest space tragedy was yet to come.

5

The Challenger Explosion

By the beginning of 1986, the space shuttles had gone into space on twenty-four successful missions. Such success created the common impression that all danger had been eliminated from manned spaceflight. To the public, the launch of a space shuttle seemed like a complicated but almost routine operation.

That false impression was tragically shattered one cold January day.

The space shuttle *Challenger* lifted off the launchpad late on the icy morning of January 28, 1986. It carried a crew of seven, including schoolteacher Christa McAuliffe. Because of McAuliffe's presence on the shuttle, many schoolchildren around the country were watching the countdown and launch on television.

Unknown to everyone, one of the solid rocket boosters began spilling its propellant through a faulty seal called an O-ring the moment the engines ignited. No one noticed the black smoke pouring from the seal as the shuttle left the pad. As far as anyone could tell, the shuttle's crew was on its way into space.

"*Challenger* now heading downrange," said Mission Control, which meant that the shuttle was heading east over the Atlantic Ocean.

Everything appeared to be going well. After the shuttle passed through the part of launch in which it has its roughest ride through the atmosphere, called max Q, the engines were ready to be pushed to even higher speed.

"*Challenger*, Go at throttle up," said Mission Control.

"Roger," commander Dick Scobee responded. "Go at throttle up."

The space shuttle Challenger *lifts off on an icy morning in January. One minute later, the mission turned into a tragedy.*

Suddenly the unimaginable happened.

"Uh oh," said pilot Mike Smith.[1]

At seventy-three seconds into the launch, *Challenger* exploded into a ball of flame and smoke. Debris was blasted outward and upward as the mass of flame and wreckage continued skyward with the exploding spacecraft's forward motion. The two solid rocket boosters went sailing out of control beyond the expanding cloud of gas and smoke. Those watching the launch from the observation area or on television did not know exactly what had happened. No one had ever seen anything like this. For thirty seconds, the public address loudspeakers at Cape Canaveral were silent.

As the people watching the launch waited for some word from the public address announcer, they realized that the column of smoke in the sky was no longer going higher. Something terrible had happened.

Finally the announcer spoke.

"Flight controllers are looking very carefully at the situation. Obviously a major malfunction." Later came more grim news. "We have a report from the flight dynamics officer that the vehicle has exploded. The flight director confirms that. We are . . . waiting for word of any recovery forces in the down-range field."[2]

The repeated images of the shuttle exploding on television were shocking. The blast was so terrible, it did not look as if anyone could have survived. The orbiter appeared to have been blown to bits by the explosion.

The Challenger *shuttle explodes seconds after liftoff. It remains one of America's worst space tragedies.*

The television networks followed the story live for the rest of the day.

At 4:30 P.M. NASA officials held a press conference. They announced that recovery forces in the Atlantic Ocean impact area had found no evidence of survivors. The report confirmed what everyone feared—the crew of *Challenger* had been killed.

The explosion of *Challenger* was America's greatest space tragedy. The public felt the tragedy perhaps even more deeply than the *Apollo 1* fire. The *Challenger* crew seemed so representative of America. With their background as military test pilots, commander Dick Scobee and pilot Mike Smith fit the classic astronaut mold. But the rest of the crew was different.

Ellison Onizuka, from Hawaii, had already become the first Japanese American to fly into space on a previous mission. Ron McNair had been the second African American to fly aboard the shuttle, and Judith Resnik was the second American woman in space.

Gregory Jarvis did not work for NASA but was an engineer for the Hughes Aircraft Corporation. He was aboard the shuttle to conduct experiments on how fluids react to various motions in space. Christa McAuliffe was a high school social studies teacher from Concord, New Hampshire. She had been selected from more than eleven thousand applicants to become the first teacher in space. It was especially easy for the public to identify with the journey into space of someone like McAuliffe.[3]

The later recovery of the crew compartment and of the astronauts' remains showed that not all of the astronauts were killed by the violent explosion. Four of their emergency air packs were recovered, and three of them had been activated. One of them was pilot Mike Smith's. Interestingly, Smith's pack had been mounted behind his cockpit seat, which means that Ellison Onizuka, who was seated behind Smith, probably activated his pack for him. More than half of the oxygen in these packs had been used, meaning that at least three astronauts had been breathing during the crew compartment's two-and-a-half minute fall to the Atlantic Ocean.

While they may have been breathing, it does not mean that they were conscious. The three astronauts who were breathing may have lost consciousness at any time if the crew compartment had been cracked or broken by the explosion. Any such crack in the crew compartment would have caused the cabin to lose pressure. At such high altitudes, this loss of pressure would have caused the crew to lose consciousness.

The Challenger *mission carried a crew of seven. From left to right are* Christa McAuliffe, Gregory Jarvis, Judith Resnik, Dick Scobee, Ron McNair, Mike Smith, *and* Ellison Onizuka. *The violent explosion of* Challenger *killed all seven astronauts.*

In any case, if any of the astronauts remained alive or conscious throughout the crew compartment's fall to the ocean, they would have immediately been killed by the compartment's two hundred-mile-per-hour impact with the water.

Like the aftermath of the *Apollo 1* fire, the investigation that followed the *Challenger* explosion found that the tragedy could have been avoided. The failure of the O-ring seal was soon identified as the cause of the disaster. But the problems went deeper than that.

The investigating commission found that NASA had not responded adequately to warnings about the O-ring seal design. The decision to launch the shuttle on that cold January morning had also been misguided. Evidence already existed that showed how the O-ring seals had suffered damage during past launches at lower temperatures. The coldest of these launches so far had been 53 degrees Fahrenheit. The temperature on the morning *Challenger* was launched was only 32 degrees Fahrenheit.

A number of management practices in the shuttle program were changed in order to make the preparation for each launch a safer operation. Two and a half years passed before NASA was ready to launch another shuttle mission. During that time, many improvements were made to the shuttle's design in an effort to make it a safer spacecraft.

On September 29, 1988, it was finally time for another launch. With the *Challenger* disaster fresh and vivid in everyone's memory, no one had forgotten the

The investigation of the Challenger *explosion revealed that extremely cold temperatures had damaged the O-ring seals.*

dangers of launching a vehicle into space.

At 11:37 A.M., the space shuttle *Discovery* lifted off the pad.

"Liftoff!" the announcer said. "Liftoff. Americans return to space as *Discovery* clears the tower."

This time the O-rings did their job. The solid rocket boosters worked perfectly, then safely separated from *Discovery*'s external tank. *Discovery* thundered onward into space. A crew of American astronauts were in space once again. Shortly after reaching orbit, commander Rick Hauck said a little thank-you to the thousands of people whose hard work had helped get the shuttle back into space.

The space shuttle successfully returns to space as Discovery *carries the first crew since the* Challenger *disaster.*

"We sure appreciate you all gettin' us up into orbit," he said, "where we should be."[4]

There are many risks and dangers in spaceflight. Since the first days of Mercury until today, the launch is still the most dangerous phase of any flight.

To put it simply, spaceflight is a dangerous business. People's lives have been lost, and more may be lost in the future. But those who fly in space believe their pursuit of knowledge about the universe is worth the risk.

Gus Grissom, who lost his life in the *Apollo 1* fire, summed it up best:

> There will be risks . . . and sooner or later, inevitably, we're going to run head-on into the law of averages and lose somebody. I hope this never happens . . . but if it does, I hope the American people won't feel it's too high a price to pay for our space program. None of us was ordered into manned spaceflight.[5]

Like Gus Grissom, all astronauts are volunteers. They volunteer to face the dangers. They take the risk of losing their lives.

To expand our scientific knowledge of ourselves and our universe, they believe it is worth it.

CHAPTER NOTES

Chapter 1. Dangerous Steps into Space

1. "Spaceflight Part 2: The Wings of Mercury," PBS Video (1985).

2. Mark Wade, *Encyclopedia Astronautica Web page,* October 1998, <http://solar.rtd.utk.edu/~mwade/chrono/19604.htm#5836>.

Chapter 2. The Launch

1. David Baker, *The History of Manned Space Flight* (New York: Crown Publishers, Inc., 1982), p. 224.

2. "Spaceflight Part 2: The Wings of Mercury," PBS Video (1985).

3. Michael Collins, *Liftoff: The Story of America's Adventure in Space* (New York: Grove Press, 1988), p. 209.

4. Ibid., p. 210.

5. Baker, p. 554.

Chapter 3. The Apollo 1 Fire

1. Alan Shepard and Deke Slayton, *Moon Shot: The Inside Story of America's Race to the Moon* (Atlanta: Turner Publishing, Inc., 1994), pp. 194–195.

2. Peter Bond, *Heroes in Space: From Gagarin to Challenger* (New York: Basil Blackwell, Inc., 1988), p. 142.

3. Shepard and Slayton, p. 197.

4. Ibid., p. 198.

5. David Baker, *The History of Manned Space Flight* (New York: Crown Publishers, Inc., 1982), p. 276.
6. Shepard and Slayton, p. 198.
7. Bond, p. 139.
8. Baker, p. 276.
9. Bond, p. 139.
10. Shepard and Slayton, p. 207.
11. Ibid., pp. 202–203.
12. Michael Collins, *Liftoff: The Story of America's Adventure in Space* (New York: Grove Press, 1988), p. 136.
13. Bond, pp. 140–142.

Chapter 4. Apollo 12
1. Peter Bond, *Heroes in Space: From Gagarin to Challenger* (New York: Basil Blackwell, Inc., 1988), p. 214.
2. Alan Shepard and Deke Slayton, *Moon Shot: The Inside Story of America's Race to the Moon* (Atlanta: Turner Publishing, Inc., 1994), p. 255.
3. Ibid.
4. David Baker, *The History of Manned Space Flight* (New York: Crown Publishers, Inc., 1982), p. 365.
5. Shepard and Slayton, pp. 255–256.
6. Bond, pp. 215–216.
7. Ibid., p. 215.

Chapter 5. The Challenger Explosion
1. Peter Bond, *Heroes in Space: From Gagarin to Challenger* (New York: Basil Blackwell, Inc., 1988), p. 439.
2. Richard S. Lewis, *Challenger: The Final Voyage* (New York: Columbia University Press, 1988), p. 21.
3. Michael Collins, *Liftoff: The Story of America's Adventure in Space* (New York: Grove Press, 1988), pp. 224–225.
4. "From Disaster to Discovery," Great TV News Stories, ABC News Video (1989).
5. Bond, p. 142.

GLOSSARY

blockhouse—A strong, thickly protected building near the launchpad where engineers and technicians monitor the launch.

capsule communicator (capcom)—The person at Mission Control who communicates directly with astronauts in the spacecraft; the capcom is usually another astronaut.

combustion—The chemical process of burning fuel, such as what occurs in a rocket engine to create thrust.

command module—The main Apollo spacecraft that carried astronauts to and from the Moon. This crew capsule carried the three astronauts and was equipped with a heat shield for reentry into Earth's atmosphere.

external fuel tank—The large tank attached to the shuttle that supplies liquid fuel to the shuttle orbiter's main engines during liftoff.

hold-down clamps—Large clamps on the launchpad that hold the rocket or space shuttle to the pad until it has reached the proper thrust for liftoff.

lunar module—The Apollo spacecraft designed to land two astronauts on the Moon. After their activities on the lunar surface, the astronauts blasted the lunar module back into lunar orbit to redock with the command module, which then carried the astronauts back to Earth.

Mission Control—The NASA control center in Houston, Texas, that maintains contact with astronauts in space, monitors spacecraft systems, and controls the activities of all manned spaceflights.

National Aeronautics and Space Administration (NASA)—The United States government agency that administers the country's space program.

orbiter—The winged spacecraft that is usually referred to as the space shuttle. The orbiter is one of the four parts of the space shuttle launch assembly.

O-rings—The O-shaped rubber ring used to seal together sections of the solid rocket boosters.

propellant—Any liquid or solid substance that is used as combustion fuel to create thrust in a rocket engine.

rendezvous—A planned meeting of two objects, such as when two spacecraft closely approach each other in orbit.

simulation—Activities in which people practice in conditions similar to those they will experience in reality. Astronauts use many different kinds of simulations to prepare for the activities they will conduct during their missions.

solid rocket booster—A rocket that uses solid explosive powders for fuel.

thrust—The force produced by the combustion energy of a rocket engine that pushes the rocket in the opposite direction of the force.

FURTHER READING

Books

Berliner, Don. *Living in Space.* Minneapolis, Minn.: The Lerner Publishing Group, 1993.

Bernards, Neal. *Mir Space Station.* Mankato, Minn.: The Creative Company, 1999.

Cole, Michael D. *Challenger: America's Space Tragedy.* Springfield, N.J.: Enslow Publishers, Inc., 1995.

Landau, Elaine. *Space Disaster.* New York: Franklin Watts, Inc., 1999.

Internet Addresses

Data Matrix, Inc. *The Astronaut Connection,* n.d. <http://nauts.com> (April 21, 1999).

Bray, Becky, and Patrick Meyer. *Kids Space.* March 27, 1998. <http://liftoff.msfc.nasa.gov/kids> (April 21, 1999).

National Aeronautics and Space Administration. "Welcome to the Future of Spaceflight." *Office of Spaceflight.* n.d. <http://www.hq.nasa.gov/osf> (April 21, 1999).

INDEX

A

Aldrin, Edwin "Buzz", 27
Apollo 1, 19–24, 36, 38
    investigation, 24, 26
Apollo 11, 27
Apollo 12, 27–32
Apollo 13, 32
Armstrong, Neil, 27
Atlas rocket, 8

B

Baykonur (Soviet launch
    facility), 5–7
Bean, Alan, 27

C

Chaffee, Roger, 17, 19, 21,
    23–24
Challenger space shuttle,
    33–38
    investigation, 38–39
Conrad, Charles, 27, 28, 30,
    31, 32
Cunningham, Walter, 26

D

Discovery space shuttle, 40

E

Eisele, Donn, 26
ejection seats, 10–11

F

Friendship 7, 8

G

Gemini 6, 10–11, 13
Gemini 7, 10, 11, 13
Glenn, John, 8
Gordon, Richard, 27, 30
Grissom, Gus, 8, 17, 19–20,
    21, 22, 23, 24, 41

H

Hauck, Rick, 40, 41

J

Jarvis, Gregory, 36–37

K

Kennedy, John F., 18
Khrushchev, Nikita, 6

M

McAuliffe, Christa, 33, 37
McNair, Ronald, 36

N

NASA, 17, 18, 24, 25, 28, 36, 39

Nixon, Richard, 28

O

Onizuka, Ellison, 36, 37

orbital maneuvering system (OMS) engines, 13, 15

orbiter, 13, 14, 15, 35

O-rings, 34, 38–39, 40

R

R-16 rocket, 5

Redstone rocket, 8

Resnik, Judith, 36

S

Saturn 1B rocket, 17, 19

Saturn V rocket, 9, 15–16, 29, 31

Schirra, Wally, 10–11, 13, 26

Scobee, Dick, 34, 36

Shepard, Alan, 8

Slayton, Deke, 22

Smith, Mike, 35, 36, 37

solid rocket boosters (SRBs), 9, 13, 14, 15, 34, 35, 40

space shuttle, 13–15, 32, 33–37, 38, 39, 40

launch assembly, 13

launch phase, 14–15

Stafford, Tom, 10, 11, 13

T

Titan rocket, 8

W

White, Ed, 17, 19, 23–24